Daniel Fretwell

Class Field Theory

GRIN Verlag

Bibliografische Information der Deutschen Nationalbibliothek:

Die Deutsche Bibliothek verzeichnet diese Publikation in der Deutschen National-
bibliografie; detaillierte bibliografische Daten sind im Internet über http://dnb.d-
nb.de/ abrufbar.

Imprint:

Copyright © 2011 GRIN Verlag GmbH
Druck und Bindung: Books on Demand GmbH, Norderstedt Germany
ISBN: 978-3-640-96929-6

This book at GRIN:

http://www.grin.com/en/e-book/175757/class-field-theory

GRIN - Your knowledge has value

Der GRIN Verlag publiziert seit 1998 wissenschaftliche Arbeiten von Studenten, Hochschullehrern und anderen Akademikern als eBook und gedrucktes Buch. Die Verlagswebsite www.grin.com ist die ideale Plattform zur Veröffentlichung von Hausarbeiten, Abschlussarbeiten, wissenschaftlichen Aufsätzen, Dissertationen und Fachbüchern.

Visit us on the internet:

http://www.grin.com/

http://www.facebook.com/grincom

http://www.twitter.com/grin_com

Class Field Theory

Daniel Fretwell

June 12, 2011

Contents

1 Introduction

One of the main tasks of 19th and 20th Century mathematics was to classify all possible finite extensions of a number field K. Even though such a classification would be infinite in nature, could there be a simple way to describe all such extensions in terms of the arithmetic of K? The answer to this question is not yet fully known although many conjectures have been made. When we restrict to studying finite extensions L/K that have Abelian Galois group, the question is solved in its entirety by class field theory. In fact if we go even further and consider unramified Abelian extensions (to be defined later) then we find a surprising correspondence with subgroups of the ideal class group of K.

Class field theory is mainly about the interplay between generalised ideal class groups and Abelian extensions. Of course, nearly every word of this will need a thorough explanation (except I assume familiarity with the notions of group and field extension). Essentially we invent a more general type of ideal class group and see the correspondence extend to cater for any Abelian extension, using a special map called the Artin map. Then finding all Abelian extensions of a number field is the same as looking at these generalised ideal class groups and their finitely many subgroups. We introduce the notion of a modulus to make this process run more smoothly.

The fact that we have a collection of correspondences between these finite groups and certin Abelian extensions implies that there exists in some sense a set of maximal Abelian extensions, with respect to the different ramification possibilities. These are the ray class fields and it is certainly not a trivial observation that such maximal Abelian extensions even exist.

Applying this to the unramified case guarantees us the existence of a maximal unramified Abelian extension, the so called Hilbert class field K_H. This field is a very important field and is useful in many situations.

1

Since class field theory is quite an in depth area of number theory, I defer proofs of the main theorems to Semester 2 and devote this project to describing the theory itself. Also in Semester 2 I approach certain applications of the theory. The project is slightly longer than guidelines, for which I apologise. I felt that there was a need to explain the material thoroughly enough to make this project enjoyable to the reader.

I assume familiarity with the basic concepts of algebraic number theory and Galois theory. In particular I use the terms number field, ring of integers, Galois group and Galois correspondence freely. For those that need an introduction to algebraic number theory and Galois theory, consult Stewart and Tall [1] or Stewart [2]. I assume knowledge up to the theory of factorisation into prime ideals, including knowledge of ramification.

2   A brief survey of algebraic number theory

We recall a few definitions and results of algebraic number theory. First we define a number field.

Definition 2.0.1. A *number field* is a field $\mathbb{Q} \subseteq K \subseteq \mathbb{C}$ such that the extension $K/\mathbb{Q}$ has finite degree, i.e. K has finite dimension as a vector space over $\mathbb{Q}$. The *degree* of a number field is the degree of the extension $K/\mathbb{Q}$.

Each element of a number field is algebraic over $\mathbb{Q}$, which is the same as being algebraic over $\mathbb{Z}$. Take the elements of a number field that satify monic polynomials over $\mathbb{Z}$. These form a ring:

Definition 2.0.2. The *ring of integers* of a number field K is the ring of elements of K that each satisfy a monic polynomial over $\mathbb{Z}$. It is denoted $\mathfrak{O}_K$.

We can assign sizes to the elements of a number field by using the Galois automorphisms with respect to a smaller number field. Measures like this are called *norms* and are helpful in comparing elements by size.

Factorisation of elements into irreducibles within the ring of integers is not always unique (although it is always possible). Remarkably we can always factorise the ideals of the ring of integers into prime ideals in a unique fashion:

Theorem 2.0.3. *Given an ideal* $\mathfrak{a}$ *of* $\mathfrak{O}_K$, *there exist prime ideals* $\mathfrak{p}_1, \mathfrak{p}_2, \ldots, \mathfrak{p}_g$ *of* $\mathfrak{O}_K$ *and positive integers* $e_1, e_2, \ldots, e_g$ *such that:*

$$\mathfrak{a} = \mathfrak{p}_1^{e_1} \mathfrak{p}_2^{e_2} \ldots \mathfrak{p}_g^{e_g}$$

Furthermore, this factorisation is unique up to ordering of the prime ideals.

Definition 2.0.4. We say that $\mathfrak{a}$ ramifies if $e_i > 1$ for some i.

We can actually define the norm of an ideal:

Definition 2.0.5. Given an ideal $\mathfrak{a}$ of $\mathfrak{O}_K$, we define its *norm* to be $N(\mathfrak{a}) = |\mathfrak{O}_K/\mathfrak{a}|$

The above norm always gives values in $\mathbb{Z}$ and is multiplicative, i.e. $N(\mathfrak{a}_1 \mathfrak{a}_2) = N(\mathfrak{a}_1) N(\mathfrak{a}_2)$ for any two ideals $\mathfrak{a}_1, \mathfrak{a}_2$ of $\mathfrak{O}_K$. Later we see how to define a more general and useful norm which is relative to other number fields. The output of this norm will be products of prime ideal powers in a certain ring of integers.

3   The ideal class group

The ideal class group is a very helpful construction in algebraic number theory and yet it is a group that still is not fully understood. In this section, the notion of a fractional ideal of a number field K is motivated and we investigate the factorisation properties of such structures. We then see that the non-zero fractional ideals form an Abelian group and construct the ideal class group from this. This always turns out to be a finite group and we call the order the class number of the number field K, denoted h_K.

The ideal class group in some sense measures the failure of unique factorisation and using simple group theoretic tools we can see how the class number helps us determine in some cases whether a given ideal is principal.

3.1 Fractional ideals

Let K be a number field and $\mathfrak{O}_K$ be its ring of integers. It is known that we have unique factorisation of any non-zero ideal of $\mathfrak{O}_K$ into prime ideals of $\mathfrak{O}_K$. We consider a larger class of objects, the fractional ideals of a number field.

Definition 3.1.1. A *fractional ideal* of a number field K is a non-zero finitely generated $\mathfrak{O}_K$-submodule of K.

Informally a fractional ideal of K is an ideal of $\mathfrak{O}_K$ divided by some element of $K^\times$. They can be shown to be of the form $c^{-1}\mathfrak{a}$ where $c \in K^\times$ and $\mathfrak{a}$ is an ideal of $\mathfrak{O}_K$. It should be noted that when explicitly working out the elements of a fractional ideal we work inside K and not just inside $\mathfrak{O}_K$, otherwise the definition would not make sense.

As stated above, we have the following:

Lemma 3.1.2. *The set of non-zero fractional ideals of K form an Abelian group under multiplication. Denote this group as I_K.*

Proof. This is really a sketch proof. The existence of an inverse is the hard part to prove and is found on p.108 of [1].

Given two non-zero fractional ideals $\mathfrak{b}_1 = c_1^{-1}\mathfrak{a}_1$ and $\mathfrak{b}_2 = c_2^{-1}\mathfrak{a}_2$ for non-zero ideals $\mathfrak{a}_1$ and $\mathfrak{a}_2$ of $\mathfrak{O}_K$, we see that the product is $\mathfrak{b}_1\mathfrak{b}_2 = (c_1 c_2)^{-1}\mathfrak{a}_1\mathfrak{a}_2$ by the commutativity of K. This is easily seen to be another non-zero fractional ideal of K since the product of ideals of $\mathfrak{O}_K$ is another ideal.

Associativity follows from associativity of multiplication inside K.

The identity element is taken to be the ordinary ideal $\mathfrak{O}_K$ and the inverse of a fractional ideal $\mathfrak{b}$ is the fractional ideal $\mathfrak{b}^{-1} = \{x \in K \mid x\mathfrak{b} \subseteq \mathfrak{O}_K\}$ $\qquad\square$

This is why the set of non-zero ideals of $\mathfrak{O}_K$ is not enough to form a group. It was impossible to define the inverse ideal without leaving $\mathfrak{O}_K$.

Given the fact that we have a group structure, we can actually form a unique factorisation into prime ideals for this bigger class of objects:

Theorem 3.1.3. *Given a non-zero fractional ideal $\mathfrak{b}$ of a number field K, there exists non-zero prime ideals $\mathfrak{p}_1, \mathfrak{p}_2, \ldots, \mathfrak{p}_s$ of $\mathfrak{O}_K$ and integers $r_1, r_2, \ldots, r_s$ such that:*

$$\mathfrak{b} = \mathfrak{p}_1^{r_1}\mathfrak{p}_2^{r_2}\ldots\mathfrak{p}_s^{r_s}$$

Futhermore, this prime ideal factorisation is unique up to ordering of prime ideal factors that have the same sign exponent.

The "fractional" nature of the factorisation should be apparent by the fact we are now allowing the exponents to take integer values and not just positive integer values.

3.2 The ideal class group

Now that we have defined the fractional ideals of a number field K, we are ready to construct the ideal class group of K. As mentioned, this group measures the extent to which unique factorisation of elements fails. The basic idea follows from the fact that unique factorisation into irreducibles holds if and only if every ideal is principal. So if we can somehow measure how far a given ideal is from being principal then we can use this information to see how much unique factorisation fails.

The best way forward is to construct the quotient group I_K/P_K, where P_K is the group of non-zero principal fractional ideals (i.e. fractional ideals of the form (α), where $\alpha \in K^\times$).

Definition 3.2.1. The group I_K/P_K is called the *ideal class group* of K, denoted Cl_K.

Surprisingly enough we have the following, non-trivial result:

Theorem 3.2.2. *The group Cl_K is a finite group.*

This tells us that every fractional ideal belongs to one of a finite number of classes of ideals that are within principal ideal multiples of each other. The order of the finite group Cl_K is called the *class number* of K, denoted h_K

There are algorithms and methods to calculate the class number and the ideal class group, but the class number has quite erratic behaviour. These algorithms involve factorising principal ideals generated by rational primes less than a specific constant and so, due to the erratic nature of the rational primes, it is not easy to implement these methods in general. Nonetheless, the methods work and can be seen as consequences of lattice theory and results of Minkowski. Explicit examples of calculating the ideal class group can be found in section 10.3 of [1].

The fact that the ideal class group is a finite group, coupled with basic group theoretical results tells us a few facts. Firstly raising any fractional ideal $\mathfrak{a}$ to the power h_K must give a principal fractional ideal, since the ideal class group has order h_K. Also if we can find an integer m coprime to h_K such that $\mathfrak{a}^m$ is a principal fractional ideal then we can immediately conclude that $\mathfrak{a}$ must itself be a principal fractional ideal. Finally, and most importantly, K has unique factorisation into irreducibles if and only if $h_K = 1$.

It is amazing that every ideal can be made principal by raising to this power h_K. Also, the second point raised above shows one of the reasons why the class number is important, since knowing it can help show when an ideal is principal in secret.

It was exactly the above that Kummer used in proving his regular case of Fermat's Last Theorem. His proof depends on deducing that the pth power of an ideal of $\mathfrak{O}_{\mathbb{Q}(\zeta_p)} = \mathbb{Z}[\zeta_p]$ being principal implies that the ideal itself is principal, when p is coprime to the class number $h_{\mathbb{Q}(\zeta_p)}$. See chapter 11 of [1] for details of this and the proof.

4 The decomposition group and Frobenius

In this section we use what we know about Galois theory to study how prime ideal factorisation works in number fields that are Galois extensions of $\mathbb{Q}$. The situation becomes much nicer in these number fields. We then investigate an important group action using the group $\mathrm{Gal}(L/K)$ where K and L are number fields. First we need to look at the Galois group of extensions concerning finite fields.

4.1 Galois theory of finite fields

It is assumed that the reader knows about finite fields. We will denote the finite field of order p^n by $\mathbb{F}_{p^n}$.

Firstly we look at extensions of the form $\mathbb{F}_{p^n}/\mathbb{F}_p$. Here we have a special kind of automorphism:

Lemma 4.1.1. *The map ϕ defined by:*

$$\phi : x \longmapsto x^p$$

is an automorphism of $\mathbb{F}_{p^n}$. Further it fixes elements of $\mathbb{F}_p$ and so is an element of $\mathrm{Gal}(\mathbb{F}_{p^n}/\mathbb{F}_p)$.

Proof. Take $x, y \in \mathbb{F}_{p^n}$.

Firstly, we expand $(x+y)^p$ using the binomial theorem. Then using the fact that the binomial coefficient $\binom{p}{i}$ is divisible by p for each $1 \leq i \leq p$ and that we are working in a field of characteristic p, we see that:

$$\phi(x + y) = (x + y)^p = x^p + y^p = \phi(x) + \phi(y)$$

so that ϕ respects addition in $\mathbb{F}_{p^n}$.

To check that ϕ respects multiplication we simply see that:

$$\phi(xy) = (xy)^p = x^p y^p = \phi(x)\phi(y)$$

by commutativity of multiplication in a field.

Finally, it is easy to check that $\phi(0) = 0$ and that $\phi(1) = 1$. Thus ϕ is an automorphism of $\mathbb{F}_{p^n}$.

The fact that ϕ fixes elements of $\mathbb{F}_p$ follows from the fact that ϕ is an automorphism, since for all non-zero $x \in \mathbb{F}_p$:

$$\phi(x) = \phi(\sum_{i=1}^{x} 1) = \sum_{i=1}^{x} \phi(1) = \sum_{i=1}^{x} 1 = x$$

The case $x = 0$ is seen above. $\qquad\square$

Definition 4.1.2. We call the map ϕ the Frobenius automorphism of $\mathbb{F}_{p^n}$, denoted Frob_p.

The Frobenius automorphsim is very important in the fact that it generates the Galois group $\mathrm{Gal}(\mathbb{F}_{p^n}/\mathbb{F}_p)$.

Theorem 4.1.3. *The Galois group* $\mathrm{Gal}(\mathbb{F}_{p^n}/\mathbb{F}_p)$ *is cyclic of degree* n, *generated by* Frob_p.

Proof. We already know that Frob_p lies in the Galois group $\mathrm{Gal}(\mathbb{F}_{p^n}/\mathbb{F}_p)$. If we can show that the elements Frob_p^k for $k = 1, 2, \ldots, n$ are all distinct then the result will follow. Equivalently we show that Frob_p^n is the identity and none of the automorphisms Frob_p^k are the identity for $k = 1, 2, \ldots, n-1$.

The first of these claims comes from the fact that the non-zero elements of a field form a group under multiplication. Thus, since $|\mathbb{F}_{p^n}^{\times}| = p^n - 1$ we must have that $x^{p^n-1} = 1$ for all $x \in \mathbb{F}_{p^n}$. With the inclusion of the element 0 we see by multiplication of both sides with x that $x^{p^n} = x$ for all $x \in \mathbb{F}_{p^n}$. Thus $\mathrm{Frob}_p^n(x) = x^{p^n} = x$ for all $x \in \mathbb{F}_{p^n}$. This could also be deduced by considering the way in which finite fields are constructed, as splitting fields for the polynomial $x^{p^n} - x$.

The second claim comes from the fact that if such a Frob_p^k was the identity then the polynomial $x^{p^k} - x$ would have to have all of the p^n elements of $\mathbb{F}_{p^n}$ as its roots. This cannot be true since the polynomial has degree $p^k < p^n$ and coupled with the fact that we are working in a field guarantees that the polynomial has at most p^k roots in total. The second claim follows. $\qquad\square$

We can now see exactly what the intermediate fields must be by using the Galois correspondence. The subgroups of the cyclic group of order n must be cyclic themselves and must have order dividing n by Lagrange's theorem. Since subgroups of a cyclic group having the same order are equal, we conclude that there must be a unique intermediate field for each divisor d of n. These intermediate fields turn out to be $\mathbb{F}_{p^d}$ for each divisor d of n.

So now we have investigated extensions of the form $\mathbb{F}_{p^n}/\mathbb{F}_p$ but what about more general extensions $\mathbb{F}_{p^n}/\mathbb{F}_{p^d}$ where $d|n$? It is a simple consequence of the Galois correspondence along with the work we have done already to see that the Galois groups of these extensions are also cyclic and are generated by Frob_p^d, the automorphism that sends x to x^{p^d}.

4.2   A Galois group action

Let $K \subseteq L$ be number fields and let $\mathfrak{p}$ be a prime ideal of $\mathfrak{O}_K$. We can extend $\mathfrak{p}$ to an ideal $\mathfrak{p}\mathfrak{O}_L$ of $\mathfrak{O}_L$. This is simply the ideal generated by the products $\alpha\beta$ for all such $\alpha \in \mathfrak{p}$ and $\beta \in \mathfrak{O}_L$. The theory of ideal factorisation guarantees a prime ideal factorisation:

$$\mathfrak{p}\mathfrak{O}_L = \mathcal{P}_1^{e_1} \mathcal{P}_2^{e_2} \ldots \mathcal{P}_g^{e_g}$$

for prime ideals $\mathcal{P}_1, \mathcal{P}_2, \ldots, \mathcal{P}_g$ of $\mathfrak{O}_L$.

Definition 4.2.1. We say that $\mathfrak{p}$ *ramifies in* L if the ideal $\mathfrak{p}\mathfrak{O}_L$ ramifies, i.e. if there exists i such that $e_i > 1$ in the above.

We now get to define the general norm described at the start of the project:

Definition 4.2.2. Define the *general norm* of a prime ideal divisor $\mathcal{P}_i$ of $\mathfrak{p}\mathfrak{O}_L$ to be $N_{L/K}(\mathcal{P}_i) = \mathfrak{p}^{f_i}$, where f_i is the degree of the finite extension of finite fields $(\mathfrak{O}_L/\mathcal{P}_i)/(\mathfrak{O}_K/\mathfrak{p})$ (this extension is considered in more detail later and is later to be proved to be what I claim). We define the general norm of a fractional ideal by extending multiplicatively in the obvious way using the factorisation in Theorem 3.1.3.

The general norm is well defined since each prime ideal of $\mathfrak{O}_L$ can be shown to divide $\mathfrak{p}\mathfrak{O}_L$ for a unique prime ideal $\mathfrak{p}$ of $\mathfrak{O}_K$. Also taking $K = \mathbb{Q}$ gives something very similar to the usual norm given to an ideal, $N(\mathcal{P}_i) = p^{f_i}$ where p is the unique rational prime contained in $\mathcal{P}_i$. We will return to study the general norm later, it will prove useful in the class field theory section.

We have the following important group action:

Theorem 4.2.3. *The group* $\mathrm{Gal}(L/K)$ *acts transitively on the set* $\{\mathcal{P}_1, \mathcal{P}_2, \ldots, \mathcal{P}_g\}$ *via:*

$$\mathcal{P}_i \longmapsto \sigma(\mathcal{P}_i) = \{\sigma(\alpha) \,|\, \alpha \in \mathcal{P}_i\}$$

Proof. First we prove that there is a group action. Take a specific $\mathcal{P}_i$ and consider $\sigma(\mathcal{P}_i)$ for any $\sigma \in \mathrm{Gal}(L/K)$. We show that this is another prime ideal of $\mathfrak{O}_L$. Suppose $ab \in \sigma(\mathcal{P}_i)$ for $a, b \in \mathfrak{O}_L$. Then $ab = \sigma(x)$ for some $x \in \mathcal{P}_i$. From this we conclude that $x = \sigma^{-1}(ab) = \sigma^{-1}(a)\,\sigma^{-1}(b) \in \mathcal{P}_i$ using the fact that σ is an automorphism. But $\mathcal{P}_i$ is a prime ideal of $\mathfrak{O}_L$ so we must have that either $\sigma^{-1}(a) \in \mathcal{P}_i$ or $\sigma^{-1}(b) \in \mathcal{P}_i$. This shows that either $a \in \sigma(\mathcal{P}_i)$ or $b \in \sigma(\mathcal{P}_i)$, thus proving that $\sigma(\mathcal{P}_i)$ is a prime ideal.

Next we show that $\sigma(\mathcal{P}_i)$ must also be a prime ideal divisor of $\mathfrak{p}\mathfrak{O}_L$. Take any $\sigma \in \mathrm{Gal}(L/K)$ and apply it to both sides of the prime ideal factorisation to get:

$$\sigma(\mathfrak{p}\mathfrak{O}_L) = \sigma(\mathcal{P}_1)^{e_1}\sigma(\mathcal{P}_2)^{e_2}\ldots\sigma(\mathcal{P}_g)^{e_g}$$

Now $\sigma(\mathfrak{p}\mathfrak{O}_L) = \sigma(\mathfrak{p})\sigma(\mathfrak{O}_L) = \mathfrak{p}\mathfrak{O}_L$. This follows from the fact that σ fixes elements of K and the fact that σ sends integers of L to integers of L (this does not necessarily mean that it fixes $\mathfrak{O}_L$).

Thus we have that:

$$\mathfrak{p}\mathfrak{O}_L = \sigma(\mathcal{P}_1)^{e_1}\sigma(\mathcal{P}_2)^{e_2}\ldots\sigma(\mathcal{P}_g)^{e_g}$$

and since we know that each $\sigma(\mathcal{P}_i)$ is also prime ideal of $\mathfrak{O}_L$, it follows by uniqueness of factorisation that $\sigma(\mathcal{P}_i)$ is also a prime ideal divisor of $\mathfrak{p}\mathfrak{O}_L$. Thus there exists a possible group action here.

The other group action axioms can easily be checked. The transitivity is the hard part of this theorem and a proof using the Chinese remainder theorem can be found on p.12 of Lang [3]. $\qquad\square$

An important corollary of this theorem is:

Corollary 4.2.4. *Given the factorisation above:*

$$\mathfrak{p}\mathfrak{O}_L = \mathcal{P}_1^{e_1}\mathcal{P}_2^{e_2}\ldots\mathcal{P}_g^{e_g}$$

inside a Galois extension L/K *of degree* n, *all of the exponents* e_i *are equal to some fixed positive integer* e. *Also for each* i, *setting the general norm as* $N_{L/K}(\mathcal{P}_i) = \mathfrak{p}^{f_i}$ *for a positive integer* f_i, *we have that all of the* f_i *values are equal to a fixed positive integer* f.

Thus the usual relationship:

$$\sum_{i=1}^{g} e_i f_i = n$$

found on p.24 of [3] becomes the much nicer relationship:

$$efg = n$$

Finally as a consequence the factorisation becomes:

$$\mathfrak{p}\mathfrak{O}_L = (\mathcal{P}_1\mathcal{P}_2\ldots\mathcal{P}_g)^e$$

with $N_{L/K}(\mathcal{P}_i) = \mathfrak{p}^f$ *for all* i.

Proof. This is an immediate consequence of the transitivity of the group action and the uniqueness of factorisation. $\qquad\square$

Now we choose one of the $\mathcal{P}_i$. By the nature of a group action, there will be a subgroup of $\mathrm{Gal}(L/K)$ consisting of automorphisms that send $\mathcal{P}_i$ to itself. This is the stabiliser of $\mathcal{P}_i$ under this group action.

Definition 4.2.5. This subgroup stabilizer of $\mathcal{P}_i$ is called the *decomposition group* of $\mathcal{P}_i$, denoted $D_{\mathcal{P}_i/\mathfrak{p}}$.

So to summarise, given a prime ideal $\mathfrak{p}$ of $\mathfrak{O}_K$ we can extend to an ideal $\mathfrak{p}\mathfrak{O}_L$ of $\mathfrak{O}_L$. The group $\mathrm{Gal}(L/K)$ then acts transitively on the prime ideal divisors $\mathcal{P}_i$ of $\mathfrak{p}\mathfrak{O}_L$. Thus there exists a subgroup $D_{\mathcal{P}_i/\mathfrak{p}}$ of $\mathrm{Gal}(L/K)$ for each $\mathcal{P}_i$ called the decomposition group. This subgroup consists of the stabiliser of $\mathcal{P}_i$ under the group action.

Theorem 4.2.6. *Let e,f,g be described as in Corollary 4.2.4. Then $D_{\mathcal{P}_i/\mathfrak{p}}$ has order ef for any choice of $\mathcal{P}_i$.*

Proof. Suppose that L/K has degree n. As always in this section we are assuming that L/K is a Galois extension, thus $\mathrm{Gal}(L/K)$ has order n.

Now given a prime ideal divisor $\mathcal{P}_i$ of $\mathfrak{p}\mathfrak{O}_L$, we know that the orbit of $\mathcal{P}_i$ under the group action has g elements since the action is transitive and by definition g is the number of prime ideal divisors. Also, by definition $D_{\mathcal{P}_i/\mathfrak{p}}$ is the stabiliser of $\mathcal{P}_i$ under the group action.

Thus by the orbit-stabiliser theorem we have that:

$$|D_{\mathcal{P}_i/\mathfrak{p}}|\, g = |\mathrm{Gal}(L/K)| = n$$

But we also have that $efg = n$ and so the result follows. $\qquad\square$

Corollary 4.2.7. *When $\mathfrak{p}$ is unramified in L, i.e. $\mathfrak{p}\mathfrak{O}_L$ is unramified as an ideal of $\mathfrak{O}_L$, then we have that $|D_{\mathcal{P}_i/\mathfrak{p}}| = f$.*

Proof. We already know that $|D_{\mathcal{P}_i/\mathfrak{p}}| = ef$ and when $\mathfrak{p}$ is unramified in L, we have that $e = 1$ so that $|D_{\mathcal{P}_i/\mathfrak{p}}| = f$. $\qquad\square$

4.3 The Artin symbol

So what is the importance of the work we have done in this section so far? Well actually we shall see in a moment that the decomposition group $D_{\mathcal{P}_i/\mathfrak{p}}$ can always be connected to the Galois group of an extension of finite fields via a group homomorphism.

However we know that the Galois group of such an extension is generated by a Frobenius automorphism. Thus it will follow that the decomposition group is generated by a coset of "Frobenius like" elements.

We then obvserve that when the prime ideal $\mathfrak{p}$ is unramified, the corresponding decomposition groups for each prime ideal divisor $\mathcal{P}_i$ will each contain a unique such element. These will be called the *Artin symbols* of the $\mathcal{P}_i$.

To get started with this process consider a finite Galois extension of number fields L/K and take a prime ideal $\mathfrak{p}$ of $\mathfrak{O}_K$. Then for each of its prime ideal divisors $\mathcal{P}_i$ in $\mathfrak{O}_L$ we have a decomposition group $D_{\mathcal{P}_i/\mathfrak{p}}$.

Fix a specific $\mathcal{P}_j$ and note that $\mathfrak{O}_L/\mathcal{P}_j$ is a finite field, since it is a finite integral domain ($\mathcal{P}_j$ is a prime ideal). Now each of the elements of $D_{\mathcal{P}_j/\mathfrak{p}}$, being K-automorphisms of L stablising $\mathcal{P}_j$, can be turned into automorphisms of $\mathfrak{O}_L/\mathcal{P}_j$ via:

$$\phi : \sigma \longmapsto \bar{\sigma}$$

where:

$$\bar{\sigma}(x + \mathcal{P}_j) = \sigma(x) + \mathcal{P}_j$$

for all $x \in \mathfrak{O}_L$. The elements $\bar{\sigma}$ are easily checked to be automorphisms of $\mathfrak{O}_L/\mathcal{P}_j$. Here it is critical that $\sigma \in D_{\mathcal{P}_j/\mathfrak{p}}$ since otherwise the induced homomorphism would be from $\mathfrak{O}_L/\mathcal{P}_j$ to $\mathfrak{O}_L/\sigma(\mathcal{P}_j)$, which is of course $\mathfrak{O}_L/\mathcal{P}_j$ exactly when $\sigma \in D_{\mathcal{P}_j/\mathfrak{p}}$.

Actually, each of these induced automorphisms $\bar{\sigma}$ of $\mathfrak{O}_L/\mathcal{P}_j$ fix elements of the finite field $\mathfrak{O}_K/\mathfrak{p}$ (again, it is a finite integral domain since $\mathfrak{p}$ is a prime ideal of $\mathfrak{O}_K$). This is down to the fact that σ fixes elements of K. Thus the map ϕ sends elements of the group $D_{\mathcal{P}_j/\mathfrak{p}}$ to elements of the group $\mathrm{Gal}((\mathfrak{O}_L/\mathcal{P}_j)/(\mathfrak{O}_K/\mathfrak{p}))$.

We have the following:

Theorem 4.3.1. *The map:*

$$\phi : D_{\mathcal{P}_j/\mathfrak{p}} \longrightarrow \mathrm{Gal}((\mathfrak{O}_L/\mathcal{P}_j)/(\mathfrak{O}_K/\mathfrak{p}))$$

described above is a surjective group homomorphism.

Proof. We must show that $\phi(\sigma\tau) = \phi(\sigma)\phi(\tau)$ for all $\sigma, \tau \in D_{\mathcal{P}_j/\mathfrak{p}}$. Now $\phi(\sigma)$ is the automorphism $\bar{\sigma}$ of $\mathfrak{O}_L/\mathcal{P}_j$ defined by:

$$\bar{\sigma}(x + \mathcal{P}_j) = \sigma(x) + \mathcal{P}_j$$

for all $x \in \mathfrak{O}_L$. We define $\phi(\tau)$ similarly.

So $\phi(\sigma)\phi(\tau)$ is the automorphism $\bar{\sigma}\bar{\tau}$ of $\mathfrak{O}_L/\mathcal{P}_j$ defined by:

$$\bar{\sigma}\bar{\tau}(x + \mathcal{P}_j) = \bar{\sigma}(\tau(x) + \mathcal{P}_j) = \sigma(\tau(x)) + \mathcal{P}_j$$

for all $x \in \mathfrak{O}_L$. This can be rewritten as:

$$(\sigma\tau)(x) + \mathcal{P}_j = \overline{\sigma\tau}(x + \mathcal{P}_j)$$

which is the automorphism of $\mathfrak{O}_L/\mathcal{P}_j$ defined by $\overline{\sigma\tau}$. But this is $\phi(\sigma\tau)$ and thus we have a group homomorphism.

Proof of the surjectivity is not so simple. It is easy to establish this when $\mathfrak{p}$ is unramified in L since then proving surjectivity is equivalent to proving injectivity of ϕ (both groups have the same finite order). A full proof of surjectivity applicable to all cases can be found on p.15 of Lang [3]. □

Definition 4.3.2. The kernel of ϕ is called the *inertia group* of $\mathcal{P}_j$ with respect to $\mathfrak{p}$, denoted $I_{\mathcal{P}_j/\mathfrak{p}}$.

Since the identity element of $\mathrm{Gal}((\mathfrak{O}_L/\mathcal{P}_j)/(\mathfrak{O}_K/\mathfrak{p}))$ is the automorphism that fixes all elements of $\mathfrak{O}_L/\mathcal{P}_j$ we see that the elements of the inertia group can be described explicitly by the elements $\sigma \in D_{\mathcal{P}_j/\mathfrak{p}}$ such that $\sigma(x) \equiv x \bmod \mathcal{P}_j$ for all $x \in \mathfrak{O}_L$.

It is now an immediate consequence of the First Isomorphism Theorem that:

$$D_{\mathcal{P}_j/\mathfrak{p}} \,/\, I_{\mathcal{P}_j/\mathfrak{p}} \cong \mathrm{Gal}((\mathfrak{O}_L/\mathcal{P}_j)/(\mathfrak{O}_K/\mathfrak{p}))$$

Now the Galois group on the right hand side is of an extension of finite fields. As mentioned earlier this group is generated by some Frobenius automorphism. Thus, there is a unique element of $D_{\mathcal{P}_j/\mathfrak{p}} \,/\, I_{\mathcal{P}_j/\mathfrak{p}}$ that corresponds to this Frobenius automorphism. In terms of elements of $D_{\mathcal{P}_j/\mathfrak{p}}$, this means that there is a coset of $I_{\mathcal{P}_j/\mathfrak{p}}$ in $D_{\mathcal{P}_j/\mathfrak{p}}$ that behave like the Frobenius automorphisms, in the generating sense.

The following theorem breaks open the entire theory, linking the inertia group to ramification:

Theorem 4.3.3. *We have that* $|I_{\mathcal{P}_j/\mathfrak{p}}| = e$, *where* e *is the exponent in the factorisation:*

$$\mathfrak{p}\mathfrak{O}_L = (\mathcal{P}_1\mathcal{P}_2\ldots\mathcal{P}_g)^e$$

Proof. The extension $(\mathfrak{O}_L/\mathcal{P}_j)/(\mathfrak{O}_K/\mathfrak{p})$ is a finite extension of finite fields and so is a Galois extension. Thus there are f elements in its Galois group, where f is the degree of this extension, which can be shown to correspond with the value of the exponent f in the general norm $N_{L/K}(\mathcal{P}_j)$. But we know that:

$$D_{\mathcal{P}_j/\mathfrak{p}} \,/\, I_{\mathcal{P}_j/\mathfrak{p}} \cong \mathrm{Gal}((\mathfrak{O}_L/\mathcal{P}_j)/(\mathfrak{O}_K/\mathfrak{p}))$$

and thus by comparing orders of both sides we see that $\frac{ef}{|I_{\mathcal{P}_j/\mathfrak{p}}|} = f$ and it follows that $|I_{\mathcal{P}_j/\mathfrak{p}}| = e$. □

Now we can say a great deal about the "Frobenius elements" when we have unramification.

Corollary 4.3.4. *If* $\mathfrak{p}$ *is unramified in* L *then we have the isomorphism:*

$$D_{\mathcal{P}_j/\mathfrak{p}} \cong \mathrm{Gal}((\mathfrak{O}_L/\mathcal{P}_j)/(\mathfrak{O}_K/\mathfrak{p}))$$

Thus there exists a unique element of $D_{\mathcal{P}_j/\mathfrak{p}}$ *that is mapped to the Frobenius automorphism of the Galois group on the right.*

Proof. This follows since if $\mathfrak{p}$ is unramified in L then $e = 1$ and so the inertia group $I_{\mathcal{P}_j/\mathfrak{p}}$ is trivial. Then the coset of $I_{\mathcal{P}_j/\mathfrak{p}}$ that is mapped to the Frobenius automorphism must consist of a single element. □

Definition 4.3.5. We call the unique element of $D_{\mathcal{P}_j/\mathfrak{p}}$ described above the *Artin symbol* of $\mathcal{P}_i$ in L/K, denoted $\left(\frac{L/K}{\mathcal{P}_i}\right)$.

The reason we denote it this way will become clear in Semester 2 when we see that this symbol really generalises the notion of Legendre symbol to higher powers. Explicitly, the Artin symbol is defined on $\mathfrak{O}_L$ by $\left(\frac{L/K}{\mathcal{P}_i}\right)(x) \equiv x^{N(\mathfrak{p})} \pmod{\mathcal{P}_i}$.

One nice property of the Artin symbol is the following:

Theorem 4.3.6. *For any $\sigma \in \operatorname{Gal}(L/K)$ we have that $\left(\frac{L/K}{\sigma(\mathcal{P}_i)}\right) = \sigma\left(\frac{L/K}{\mathcal{P}_i}\right)\sigma^{-1}$.*

Proof. This is quite a straightforward proof. Take $\sigma \in \operatorname{Gal}(L/K)$. It is easy to see that:

$$\left(\frac{L/K}{\mathcal{P}_i}\right)(\sigma^{-1}(x)) \equiv \sigma^{-1}(x^{N(\mathfrak{p})}) \pmod{\mathcal{P}_i}$$

for all $x \in \mathfrak{O}_L$.

Applying σ to both sides gives:

$$\sigma\left(\frac{L/K}{\mathcal{P}_i}\right)\sigma^{-1}(x) \equiv x^{N(\mathfrak{p})} \pmod{\sigma(\mathcal{P}_i)}$$

But we also have that

$$\left(\frac{L/K}{\sigma(\mathcal{P}_i)}\right)(x) \equiv x^{N(\mathfrak{p})} \pmod{\sigma(\mathcal{P}_i)}$$

for all $x \in \mathfrak{O}_L$.

By the uniqueness of the Artin symbol (mentioned above), it must be the case that:

$$\left(\frac{L/K}{\sigma(\mathcal{P}_i)}\right) = \sigma\left(\frac{L/K}{\mathcal{P}_i}\right)\sigma^{-1}$$

$\square$

5 Class field theory

Now we are ready to see class field theory in action. First we see how it works in a specific case, when we have an unramified Abelian extension.

Later we see how class field theory works for more general Abelian extensions after the introduction of a modulus, a formal object which will allow us to group together the ramification in such a way that we can in some sense "ignore" it.

It should be stressed that none of the main theorems of class field theory will be proved here. The proofs are extremely complicated and would not fit in this project. I approach the proofs in the Semester 2 project. Also, any field extension from now on is assumed to be a finite extension.

5.1 Class field theory in unramified abelian extensions

Now that we have defined the Artin symbol, we study its properties. These properties will lead to the main results of class field theory.

It is easy to see from Theorem 4.3.6 that when the Galois group is Abelian, we have that $\left(\frac{L/K}{\sigma(\mathcal{P}_i)}\right) = \left(\frac{L/K}{\mathcal{P}_i}\right)$ for all $\sigma \in \operatorname{Gal}(L/K)$. This means that in this case we have the nice situation where all of the Artin symbols are the same irrespective of which $\mathcal{P}_i$ is chosen. In other words an Abelian Galois group means that the Artin symbol just depends on the original prime ideal $\mathfrak{p}$ of $\mathfrak{O}_K$. This allows us to simply define the Artin symbol as $\left(\frac{L/K}{\mathfrak{p}}\right)$ here.

Definition 5.1.1. A Galois extension L/K is called an *Abelian extension* if it has Abelian Galois group.

Now we place another condition on the type of extension. For the purposes of this section we work in unramified Abelian extensions. I will try to explain this notion.

The main problem with the Artin symbol is that it is only well-defined for unramified prime ideals. It would make it easiest if we work in an extension where every prime ideal $\mathfrak{p}$ of $\mathfrak{O}_K$ is unramified in $\mathfrak{O}_L$. In this case we can always define the Artin symbol, no matter what $\mathfrak{p}$ and $\mathcal{P}_i$ we are specifying. This is what is meant to be captured by the notion of an *unramified extension*. Actually, this is not the entire definition since there is a theory of ramification for the embeddings of K into $\mathbb{R}$ too. A real embedding of K (i.e. a monomorphism from $K \longrightarrow \mathbb{R}$) is said to ramify if it can be extended to a complex embedding of L (i.e. a monomorphism from $L \longrightarrow \mathbb{C}$ that is not a real embedding).

For those that are puzzled as to why we have to consider this new kind of ramification, see [5]. The main idea is that the valuations (or absolute values) on a number field can be separated into equivalence classes, called places. Now there are two kinds of non-trivial valuation, Archimedean and non-Archimedean. The equivalence classes of non-Archimedean valuations make what are called finite places of a number field, the equivalence classes of Archimedean valuations make infinite places.

We can go further. The finite places turn out to correspond with the so called $\mathfrak{p}$-adic valuations and so are in correspondence with prime ideals of the number field. The infinite places correspond with both embeddings of K into $\mathbb{R}$ and pairs of conjugate embeddings of K into $\mathbb{C}$ (that are not real embeddings).

So the definition of ramification we use in classical algebraic number theory is really a property of valuations in disguise. In the same way, the website referenced above establishes a theory of ramification for the infinite places too. In fact only the real embeddings can ramify under this and hence we recover the definition of ramification for real embeddings I gave above.

Definition 5.1.2. A Galois extension L/K is called an *unramified extension* if we have that each of the places of K, finite or infinite, ramify in L.

So to summarise, when the Galois extension is Abelian and unramified we have a specific element $\left(\frac{L/K}{\mathfrak{p}}\right)$ of $\mathrm{Gal}(L/K)$ for each prime ideal $\mathfrak{p}$ of $\mathfrak{O}_K$. In later sections we see how to loosen the unramified assumption on our extension without changing the situation entirely. All of the results we see in this section will actually follow from the more general setting. For the rest of this section we assume that L/K is an unramified Abelian extension.

We now extend the definition of Artin symbol to the group I_K that we studied earlier. Given a non-zero fractional ideal $\mathfrak{a} \in I_K$, we know that it has a prime ideal factorisation in K of the form:

$$\mathfrak{a} = \mathfrak{p}_1^{r_1}\mathfrak{p}_2^{r_2}\ldots\mathfrak{p}_k^{r_k}$$

where $r_1, r_2, \ldots, r_k \in \mathbb{Z}$ and $\mathfrak{p}_1, \mathfrak{p}_2, \ldots, \mathfrak{p}_k$ are prime ideals of $\mathfrak{O}_K$. This is what we saw earlier.

Definition 5.1.3. We define the Artin symbol of $\mathfrak{a}$ in L/K to be:

$$\left(\frac{L/K}{\mathfrak{a}}\right) = \prod_{i=1}^{k}\left(\frac{L/K}{\mathfrak{p}_i}\right)^{r_i}$$

This is well defined since we are working with elements of a group (which have inverses) and also each of the Artin symbols on the right are well defined.

Theorem 5.1.4. *Defining the Artin symbol like this creates a group homomorphism:*

$$\left(\frac{L/K}{\cdot}\right) : I_K \longrightarrow \mathrm{Gal}(L/K)$$

$$\mathfrak{a} \longmapsto \left(\frac{L/K}{\mathfrak{a}}\right)$$

Proof. It is easily checked that:

$$\left(\frac{L/K}{\mathfrak{a}\mathfrak{b}}\right) = \left(\frac{L/K}{\mathfrak{a}}\right)\left(\frac{L/K}{\mathfrak{b}}\right)$$

for $\mathfrak{a}, \mathfrak{b} \in I_K$. This follows by considering prime ideal factorisations of $\mathfrak{a}, \mathfrak{b}$ and using the fact that the Galois group is Abelian. $\quad\square$

Definition 5.1.5. The group homomorphism above is called the *Artin map*.

The Artin map is the most important map as far as class field theory is concerned. This map provides the links we need to describe connections between Abelian extensions and generalised ideal class groups (to be defined later).

The following result will not be proved here but will be a consequence of the existence theorem in the next subsection:

Theorem 5.1.6. *Given a number field K there exists a field K_H such that K_H is a maximal unramified Abelian extension of K. This is maximal in the sense that every unramified Abelian extension of K is contained in K_H.*

Definition 5.1.7. The field K_H is called the *Hilbert class field* of K

The following will follow from the more general results in the next subsection:

Theorem 5.1.8. *The Artin map with respect to the Galois extension K_H/K is a surjection with kernel P_K, the set of non-zero principal fractional ideals inside K.*

We do not prove this here. Instead the proof is delayed until Semester 2. In fact none of the main results in this section will be proved until Semester 2, the unramified versions will follow from the more general versions in the next subsection.

Armed with Theorem 5.1.8 we then find the following corollary:

Corollary 5.1.9. *We have the following isomorphism:*

$$Cl_K = I_K/P_K \cong Gal(K_H/K)$$

where Cl_K is the ideal class group of K

Proof. By Theorem 5.1.8 we see that the Artin map $\left(\frac{K_H/K}{\cdot}\right)$ is a surjection into $Gal(K_H/K)$ with kernel P_K. Thus the result follows by the First Isomorphism Theorem. $\quad\square$

Since we have an isomorphism with the Galois group we can use the Galois correspondence to prove the following:

Theorem 5.1.10. *Let K be any number field. There is a one-to-one correspondence between subgroups of the ideal class group Cl_K and the unramified Abelian extensions M of K.*

Proof. Using Corollary 5.1.9 we see that the subgroups of the finite group Cl_K are in one to one correspondence with the subgroups of $Gal(K_H/K)$, which are in one to one correspondence with the intermediate fields $K \subseteq M \subseteq K_H$. We now check that M/K must be an unramified Abelian extension for each such M.

Firstly we note that the extension M/K must be a Galois extension. Separability is automatic and normality follows from the fact that $Gal(K_H/M)$, being a subgroup of $Gal(K_H/K)$, must be Abelian and is hence a normal subgroup.

The fact that M/K is an Abelian extension follows from the fact that $Gal(M/K) = \frac{Gal(K_H/K)}{Gal(K_H/M)}$, being the quotient of an Abelian group, must be Abelian.

Now it remains to prove that M/K is an unramified extension and remember that this is in both the prime ideal and the real embedding sense.

Take any prime ideal $\mathfrak{p}$ of K. Since K_H/K is an unramified extension we know that $\mathfrak{p}\mathfrak{O}_{K_H}$ is not divisible by the square of any prime ideal of $\mathfrak{O}_{K_H}$. Now suppose that there was a prime ideal $\mathfrak{q}$ of $\mathfrak{O}_M$ whose square divided $\mathfrak{p}\mathfrak{O}_M$. Then by factorising $\mathfrak{q}\mathfrak{O}_{K_H}$ in $\mathfrak{O}_{K_H}$ we would find at least one prime ideal divisor $\mathcal{P}$. Then it follows that the square of $\mathcal{P}$ divides $\mathfrak{p}\mathfrak{O}_{K_H}$ by simply extending the factorisation of $\mathfrak{p}\mathfrak{O}_M$ to $\mathfrak{p}\mathfrak{O}_{K_H}$. This is a contradiction and so no prime ideals of K ramify in M.

Now we only need to prove that the real embeddings of K do not ramify in M. But this is merely a consequence of extending monomorphisms. If there exists a real embedding σ of K that ramifies in M then σ can be extended to a pair of strictly complex conjugate embeddings $\tau, \bar{\tau}$ of M (i.e. not real embeddings). Then both of $\tau, \bar{\tau}$ extend to strictly complex embeddings of K_H. But this means that σ ramifies in K_H which is impossible since K_H/K is an unramified extension. Thus no such embedding exists.

This completes the proof that M/K is a unramified extension. $\hfill\square$

This result is essentially the justification for why this branch of mathematics is called class field theory. Later we will see this correspondence appearing in more general settings. The importance of this theorem is that we now see that the unramified Abelian extensions of K are all solely dependent on the ideal arithmetic inside K.

The one to one correspondence mentioned above is a consequence of the Galois correspondence, thus it is also an order reversing correspondence, i.e. big subgroups of Cl_K correspond to unramified Abelian extensions of K of small degree.

The fact that we have this corresponence coupled with the fact that Cl_K is a finite group tells us that there can only be a finite number of unramified Abelian extensions.

The Hilbert class field K_H has many interesting properties which shall be investigated in Semester 2. It is especially useful in answering the question, "Given a positive integer n, which rational primes p can be written in the form $p = x^2 + ny^2$ where x and y are integers?" It is not difficult to see that this question is closely related to the splitting of p in the ring of integers of $\mathbb{Q}(\sqrt{-n})$. To see this we note that if x and y exist for a given p then the factorisation:

$$p = (x + y\sqrt{-n})(x - y\sqrt{-n})$$

implies that the ideal $p\mathfrak{O}_{\mathbb{Q}(\sqrt{-n})}$ splits in $\mathfrak{O}_{\mathbb{Q}(\sqrt{-n})}$.

5.2 Loosening the unramified condition

The work in the last section, although useful, is open to generalisation. How can we reshape the theory to study all Abelian extensions of K, ramified or not?

The idea is to create the idea of a modulus. Basically, given an extension L/K of number fields, we are going to in some sense "throw away" the ramification of K in L and study the rest of the structure. This turns out to still be enough to give us what we want, for example we can then still define Artin symbols and Artin maps based on information within K.

Definition 5.2.1. A *modulus* $\mathfrak{m}$ of a number field K is a formal product of finitely many prime ideals of $\mathfrak{O}_K$ (not necessariliy distinct) and distinct real embeddings of K. A modulus can be written as $\mathfrak{m} = \mathfrak{m}_0\mathfrak{m}_\infty$. The *finite part* $\mathfrak{m}_0$ consists of the ideal portion of $\mathfrak{m}$ and the *infinite part* $\mathfrak{m}_\infty$ consists of the real embeddings of $\mathfrak{m}$.

So a modulus is a kind of list of "primes" of K (excluding strictly complex embeddings). The basic idea is that we will form a modulus from all of the ramified primes and then throw away the things that share factors with the modulus, leaving only the unramified information behind.

The group I_K studied earlier now contains a subgroup for each modulus $\mathfrak{m}$ of K:

Lemma 5.2.2. *Given a modulus $\mathfrak{m}$ of K, the set:*

$$I_K(\mathfrak{m}) = \{\mathfrak{a} \in I_K \mid \mathfrak{p} \nmid \mathfrak{a} \text{ for any prime ideal } \mathfrak{p} \text{ contained in } \mathfrak{m}\}$$

is a group under multiplication.

Proof. This is straightforward to check using subgroup criterion. The result should be clear intuitively. $\hfill\square$

Now that this group is defined we can recover the notion of Artin map for a general Abelian extension L/K, however it will now depend on the modulus we use. Also, for this map to be defined on the whole of $I_K(\mathfrak{m})$ we have to make the modulus $\mathfrak{m}$ big enough to contain all ramified primes of K in L since the Artin symbol is only well defined when we have unramification.

The following definition will make things easier to state:

Definition 5.2.3. Let L/K be an extension of number fields. We call a modulus $\mathfrak{m}$ of K a *complete modulus* for L/K if $\mathfrak{m}$ contains all primes of K that ramify in L (including the real infinite primes, i.e. real embeddings that ramify in L).

Definition 5.2.4. Let L/K be a finite Abelian extension of number fields and let $\mathfrak{m}$ be any complete modulus of L/K. Then we can define the Artin map:

$$\Phi_{L/K,\mathfrak{m}} : I_K(\mathfrak{m}) \longrightarrow \mathrm{Gal}(L/K)$$

The map is defined in the same way as in the previous section using Artin symbols, except that now it is only defined on the subgroup $I_K(\mathfrak{m})$ of I_K rather than the whole of I_K.

Note that this map will be different if we change to a different complete modulus but the idea is that we have the existence of such a map for each complete modulus.

Now that we have defined the Artin map for any finite Abelian extension coupled with a complete modulus, we can study the properties of the Artin map. Since we are working in an Abelian extension, it is clear that the Artin map is once again a group homomorphism for each complete modulus.

What more can be said about the Artin map and its kernel? In the unramified case we had the kernel always containing the group of principal fractional ideals P_K so surely a natural choice would be to study $P_K(\mathfrak{m})$, the group of principal ideals coprime to $\mathfrak{m}_0$?

Unfortunately this group is too big for our liking and studying a few examples will convince the reader of this (in some cases we have that the kernel is contained within this group). The more natural group we need to use is a slight modification of $P_K(\mathfrak{m})$.

Let $P_{K,1}(\mathfrak{m})$ be the set of principal fractional ideals $\alpha\mathfrak{O}_K$ of K, coprime to $\mathfrak{m}_0$, with α satisfying the added conditions:

$$\alpha \equiv 1 \bmod \mathfrak{m}_0 \quad \text{and} \quad \sigma(\alpha) > 0$$

for each real embedding σ appearing in $\mathfrak{m}$.

The following can be checked:

Lemma 5.2.5. *The set $P_{K,1}(\mathfrak{m})$ is a subgroup of $P_K(\mathfrak{m})$ for each modulus $\mathfrak{m}$ of K.*

Proof. This is a trivial thing to check. I will just show that we have closure of multiplication in this set. Take $\alpha\mathfrak{O}_K, \beta\mathfrak{O}_K \in P_{K,1}(\mathfrak{m})$, then $(\alpha\mathfrak{O}_K)(\beta\mathfrak{O}_K) = (\alpha\beta)\mathfrak{O}_K$ in $P_K(\mathfrak{m})$. We now need to show that this actually lies in $P_{K,1}(\mathfrak{m})$.

But this is obvious since we have that $\alpha\beta \equiv (1)(1) \equiv 1 \bmod \mathfrak{m}_0$ and also that $\sigma(\alpha\beta) = \sigma(\alpha)\sigma(\beta) > 0$ for all real embeddings σ in $\mathfrak{m}$. $\qquad\square$

Now that we understand the special subgroup that is going to make the theorems ahead work, we make the following definitions to make later results easier to state:

Definition 5.2.6. Let $\mathfrak{m}$ be a modulus for K. Any group H such that $P_{K,1}(\mathfrak{m}) \subseteq H \subseteq I_K(\mathfrak{m})$ is called a *congruence subgroup* for $\mathfrak{m}$ and the quotient group $I_K(\mathfrak{m})/H$ is known as a *generalised ideal class group* for $\mathfrak{m}$.

The following is a major theorem in class field theory that answers the questions raised above about the Artin map and its kernel (proof delayed until Semester 2):

Theorem 5.2.7. *(Artin reciprocity theorem) Let L/K be an Abelian extension of number fields. Then we have that:*

    1. The Artin map is a surjection for any complete modulus $\mathfrak{m}$ of L/K.

2. *If the exponents of the finite primes in the complete modulus* $\mathfrak{m}$ *are big enough then we have that* $\ker(\Phi_{L/K,\mathfrak{m}})$ *is a congruence subgroup for* $\mathfrak{m}$, *thus giving the isomorphism:*

$$I_K(\mathfrak{m})/\ker(\Phi_{L/K,\mathfrak{m}}) \cong Gal(L/K)$$

implying that Gal(L/K) *is itself isomorphic to a generalised ideal class group for some modulus* $\mathfrak{m}$.

Now actually this is not the entire Artin reciprocity theorem and the reader might be tempted to ask a few questions at this point such as; What makes this a reciprocity theorem? Where did $P_{K,1}(\mathfrak{m})$ spring from and why is this the group to study? What is so special about being a congruence subgroup?

Fortunately the rest of the Artin reciprocity theorem, along with the other main theorems of class field theory will answer these questions.

We first return to investigating the general norm defined earlier, but this time we gain more by defining it on the group $I_L(\mathfrak{m})$ rather than on the whole of I_K. It can be checked that:

Lemma 5.2.8. *The generalised norm map:*

$$N_{L/K}(.) : I_L(\mathfrak{m}) \longmapsto I_K(\mathfrak{m})$$

is a group homomorphism, hence the image of the map, $N_{L/K}(I_L(\mathfrak{m}))$ *is a subgroup of* $I_K(\mathfrak{m})$.

Artin originally went further than Part 2 of Theorem 5.2.7, to actually find the kernel of the Artin map explicitly for a given complete modulus $\mathfrak{m}$ of L/K. This is what he found:

Theorem 5.2.9. *Let* L/K *be an Abelian extension. Then for any complete modulus* $\mathfrak{m}$ *of* L/K *that is big enough to make* $\ker(\Phi_{L/K,\mathfrak{m}})$ *a congruence subgroup for* $\mathfrak{m}$, *we have that* $\ker(\Phi_{L/K,\mathfrak{m}}) = P_{K,1}(\mathfrak{m})N_{L/K}(I_L(\mathfrak{m}))$. *This then gives us the isomorphism:*

$$I_K(\mathfrak{m})/P_{K,1}(\mathfrak{m})N_{L/K}(I_L(\mathfrak{m})) \cong \text{Gal}(L/K)$$

This result explains why the group $P_{K,1}(\mathfrak{m})$ is so important in the theory, because it actually features in the kernel of the Artin map. We will delay the proof until Semester 2 since it is very difficult to tackle with our current knowledge. Theorems 5.2.7 and 5.2.9 combined tell us lots about the Artin map but it does come with problems. Firstly it only guarantees the existence of a complete modulus that makes things work smoothly but it does not give any indication as to how big the exponents in the complete modulus have to be for this to happen. This is a delicate question and will not be studied here. Secondly, there are an infinite number of complete moduli that make this theorem work, once we have found one that is "big" enough in the sense of part 2 in the above, we can make it bigger and find another modulus that makes the theory work.

What we need is to find the best possible choice, the most efficient. This is what is known as the *conductor* $\mathfrak{f}$ and its existence is guaranteed by the following theorem:

Theorem 5.2.10. *(Existence of a conductor) Let* L/K *be an Abelian extension. Out of all the complete moduli* $\mathfrak{m}$ *that create congruence subgroups, there exists exactly one such complete modulus* $\mathfrak{f}$ *with the following properties:*

1. *A prime of* K, *finite or infinite, ramifies in* L *if and only if it is contained in* $\mathfrak{f}$.

2. *Any complete modulus* $\mathfrak{m}$ *that creates a congruence subgroup is such that* $\mathfrak{f}|\mathfrak{m}$, *in the obvious way.*

So informally the conductor is a complete modulus containing exactly the ramified primes of K in L and the exponents of these ramified primes are the smallest that make the kernel of the Artin map into a congruence subgroup.

Now we turn the situation around and establish the other side of the correspondence. Starting with a modulus and finding the congruence subgroups for that modulus, do we get all of the Abelian extensions of K? The answer is yes and this is why being a congruence subgroup is an important thing.

Theorem 5.2.11. *(The Existence Theorem) Let $\mathfrak{m}$ be a fixed modulus of a number field K. Then for each congruence subgroup H of $\mathfrak{m}$ there exists a unique Abelian extension L/K such that $H = \ker(\Phi_{L/K,\mathfrak{m}})$. These Abelian extensions are such that $\mathfrak{m}$ contains all ramified primes of K in L. Thus for each congruence subgroup H of $\mathfrak{m}$, there exists a number field L such that L/K is Abelian and*

$$I_K(\mathfrak{m})/H \cong \mathrm{Gal}(L/K)$$

I should probably recap what we have found intuitively in the last few main theorems. Basically we see that we have a correspondence between generalised ideal class groups and Abelian extensions. The Artin map is the key to this correspondence.

On one hand, given an Abelian extension there exists a modulus that makes the kernel of the Artin map a congruence subgroup, thus making the Galois group of the Abelian extension into a generalised ideal class group. On the other hand we see that each Abelian extension actually comes from a specific congruence subgroup of a specific modulus. In essence we choose the modulus in order to "choose the ramification" that we want our Abelian extension to have.

Now we ask how the unramified version of class field theory fits in from this subsection. Well essentially we just take the modulus $\mathfrak{m} = (1)$. It then follows that $I_K(\mathfrak{m}) = I_K$ and $P_{K,1}(\mathfrak{m}) = P_K$, making the congruence subgroups be the groups H such that $P_K \subseteq H \subseteq I_K$. It then follows from the Artin reciprocity theorem and the Existence theorem that the Artin map is a surjection and the one to one correspondence follows from the one to one correspondence we have in this subsection.

One other nice use of the Existence theorem is that it guarantees, for each modulus $\mathfrak{m}$, the existence of a maximal Abelian extension of K. This follows since $H = P_{K,1}(\mathfrak{m})$ is itself a congruence subgroup, meaning that there exists some Abelian extension $K \subseteq K_{\mathfrak{m}}$ such that $I_K(\mathfrak{m})/P_{K,1}(\mathfrak{m}) \cong Gal(K_{\mathfrak{m}}/K)$. Further, there can be no higher Abelian extension containing this one since H is the smallest possible congruence subgroup.

Definition 5.2.12. The field $K_{\mathfrak{m}}$ is called the *ray class field* for modulus $\mathfrak{m}$.

So the ray class field is the maximal abelian extension with respect to the modulus we are using. There is no "bigger" Abelian extension containing it. If we take again the modulus $\mathfrak{m} = (1)$ then we get out the Hilbert class field from earlier.

We finish with an example of what we have encountered and pause to prove a famous theorem. Let us consider the cyclotomic extension $\mathbb{Q}(\zeta_m)/\mathbb{Q}$. The only primes of $\mathfrak{O}_{\mathbb{Q}} = \mathbb{Z}$ to ramify in $\mathbb{Q}(\zeta_m)$ are the ones dividing m (by consideration of the discriminant). With a slight abuse of notation we take our modulus to be $\mathfrak{m} = (m)\infty$, where ∞ denotes the only real embedding of $\mathbb{Q}$. The group $I_{\mathbb{Q}}(\mathfrak{m})$ is then the group of fractional ideals of $\mathbb{Q}$ of the form $\left(\frac{a}{b}\right)$ where a, b, m are all pairwise coprime and $\frac{a}{b} > 0$ (ideals generated by associates are equal, meaning this positivity can be assumed). This gives us the Artin map as follows (which we know is a surjection by the Artin reciprocity theorem):

$$\Phi_{\mathbb{Q}(\zeta_m)/\mathbb{Q},(m)\infty} : I_K(\mathfrak{m}) \longrightarrow \mathrm{Gal}(\mathbb{Q}(\zeta_m)/\mathbb{Q})$$

Let us work out the kernel of this Artin map from first principles. First, by definition of the Artin map (and replacing the Artin map with Φ for ease of writing):

$$\Phi\left(\frac{a}{b}\right) = \Phi(a)(\Phi(b))^{-1}$$

We need a small result:

Lemma 5.2.13. *We have for each $c \in \mathbb{Z}$:*

$$\Phi(c) = \sigma_c$$

where σ_n is the automorphism defined by $\zeta \longmapsto \zeta^n$.

Proof. Assume c is positive (the negative version will have the same proof). Let c have prime factorisation in $\mathbb{Z}$ given by $c = p_1^{k_1} p_2^{k_2} \ldots p_j^{k_j}$. Thus $(c) = (p_1)^{k_1}(p_2)^{k_2} \ldots (p_j)^{k_j}$ and using the fact that Φ is a homomorphism, we have that:

$$\Phi(c) = \Phi(p_1)^{k_1}\Phi(p_2)^{k_2} \ldots \Phi(p_j)^{k_j}$$

15

and so we only need to consider $\Phi(p)$ where p is a rational prime, i.e. (p) is a prime ideal of $\mathbb{Z}$.

But we know exactly how the Artin symbol acts on prime ideals. Since $N((p)) = |\mathbb{Z}/p\mathbb{Z}| = p$ in $\mathbb{Q}$ we have that:

$$\Phi(p)(x) \equiv x^p \bmod (p)$$

for all $x \in \mathfrak{O}_{\mathbb{Q}(\zeta_m)} = \mathbb{Z}[\zeta_m]$. Specifically, taking $x = \zeta_m$ we see that $\Phi(p) = \sigma_p$ (mth roots of unity are distinct mod p).

Thus, returning to the above:

$$\Phi(c) = \sigma_{p_1}^{k_1}\sigma_{p_2}^{k_2}\dots\sigma_{p_j}^{k_j} = \sigma_{p_1^{k_1}}\sigma_{p_2^{k_2}}\dots\sigma_{p_j^{k_j}} = \sigma_{p_1^{k_1}p_2^{k_2}\dots p_j^{k_j}} = \sigma_c$$

$\square$

Thus using this lemma we see that:

$$\Phi\left(\frac{a}{b}\right) = \sigma_a(\sigma_b)^{-1} = \sigma_{ab^{-1}}$$

and so $\left(\frac{a}{b}\right) \in \ker(\Phi)$ if and only if $ab^{-1} \equiv 1 \bmod m$ and $\frac{a}{b} > 0$.

Thus here:

$$\ker(\Phi) = \left\{ \left(\frac{a}{b}\right) \in I_{\mathbb{Q}}(\mathfrak{m}) \,|\, a, b, m \text{ pairwise coprime}, ab^{-1} \equiv 1 \bmod m \text{ and } \frac{a}{b} > 0 \right\}$$

This is $P_{\mathbb{Q},1}((m)\infty)$ and so $\ker(\phi) = P_{\mathbb{Q},1}((m)\infty)$ in this case.

This fact tells us that $\mathbb{Q}(\zeta_m)$ is the ray class field for $\mathbb{Q}$ with respect to the modulus $(m)\infty$. Explicitly we have the following isomorphism:

$$I_{\mathbb{Q}}((m)\infty)/P_{\mathbb{Q},1}((m)\infty) \cong \mathrm{Gal}(\mathbb{Q}(\zeta_m)/\mathbb{Q}) \cong (\mathbb{Z}/m\mathbb{Z})^{\times}$$

Actually all of the ray class fields of $\mathbb{Q}$ are either cyclotomic or are real subfields of a cyclotomic field. This follows from the fact that any Abelian extension of $\mathbb{Q}$ has its conductor either being the modulus $(m)\infty$ or (m), for some positive integer m. We have just seen this explicitly in the case of the modulus $(m)\infty$ and so no modulus with higher exponents can make a larger Abelian extension than the cyclotomic ones. Checking the case of the modulus (m) for each m always reveals real ray class fields of the form $\mathbb{Q}(\zeta_m + \zeta_m^{-1})$. These are certainly contained within $\mathbb{Q}(\zeta_m)$.

None of these deductions are easy to see straight off but after some thought and experimentation they can all be shown to be true.

We can now prove the following theorem:

Theorem 5.2.14. *(Kronecker-Weber theorem) Each Abelian extension of $\mathbb{Q}$ lies inside a cyclotomic extension $\mathbb{Q}(\zeta_m)$ for some positive integer m.*

Proof. This follows from discussion above. Each ray class field of $\mathbb{Q}$ is either a cyclotomic field $\mathbb{Q}(\zeta_m)$ or a real subfield of $\mathbb{Q}(\zeta_m)$ for some positive integer m. Therefore any Abelian extension of $\mathbb{Q}$ must be contained inside $\mathbb{Q}(\zeta_m)$ for some positive integer m. $\square$

A nice example of where this makes sense is with the special result that for rational prime p, the field $\mathbb{Q}(\zeta_p)$ always contains the quadratic subfield $\mathbb{Q}\left(\sqrt{(-1)^{\frac{p-1}{2}}p}\right)$. Considering the quadratic field itself, which must be an Abelian extension of $\mathbb{Q}$ since it has galois group cyclic of order 2, the Kronecker-Weber theorem predicts that it must lie inside some cyclotomic extension. This one can then be verified to be $\mathbb{Q}(\zeta_p)$.

The existence of the quadratic subfield above is an important result, it helps to form a proof on quadratic reciprocity and also provides some help towards results in the construction of polygons.

References

[1] I. Stewart and D. Tall, *Algebraic Number Theory and Fermat's Last Theorem – 3rd edition,* A.K.Peters, 2002.

[2] I. Stewart, *Galois Theory – 3rd edition,* Chapman Hall/CRC, 2003.

[3] S. Lang, *Algebraic Number Theory – 2nd edition,* Springer, 1994.

[4] D. A. Cox, *Primes of the Form $x^2 + ny^2$,* Wiley, 1989.

[5] *Ramification of Archimedian Places:*

http://myyn.org/m/article/ramification-of-archimedean-places